OET Hints for the Writing Subtest for Nurses Book 2

by Virginia Allum

© 'OET Hints for the Writing Subtest for Nurses Book 2'.
by Virginia Allum 2015. All Rights Reserved

Contents

The Occupational English Test for the writing subtest	5
What kind of the Tasks are used in the Sub-test?	10
What Skills are used in the Writing Sub-test?	12
Typical Assessment Types for Medical Professionals	22
How is the OET marked and assessed?	24
What Level of English are You at?	27
How Important is the Use of Good Grammar?	32
Use of Formal and Informal Verbs	37
Verb and Preposition Pairs	38
Agreement of Verbs	40
Parallel Sentences	42
Become Familiar with Common Drugs	44
Practice Examples: Peripheral Artery Disease	46
School Nurse Letters: Letter to a Parent	51
OET Writing for Nurses: Teenager with Anaemia	58
Case Notes: Deliberate Self Poisoning	63

The OET Writing for Nurses Sub-test

Many people prefer the OET to the IELTS because they already have the background knowledge in nursing in their own language.

This is an advantage, however, you have to remember at all times that the OET tests your language skills. It tests your written communication skills.

The most important skills you need to do the referral letter are:
*ability to pick out only the relevant points of the stimulus material
* ability to write a clear and concise letter of no more than 200 words.
* the use of appropriate language for the receiver of the letter. This means that you have to choose the correct level of language (medical terms or everyday terms).

It is essential that you know how to set out the letter correctly. This includes the correct layout of a letter. Remember that these are easy marks – you have all the information in the stimulus material (with the correct spelling!).

Many students make mistakes with the setting out of the address and salutation sections. This is a pity as they are easy marks.

Be careful to spell any terms which are in the stimulus material correctly. It does not look good to misspell words you have been given!

The reading time allows you to read the stimulus material, but not take notes. This is quite challenging as it means that you have to keep the location of information in your memory. And, you have to keep checking on the relevance of the information as you read it.

You have 40 minutes to write your letter. Make sure you quickly plan out the paragraphs in the rough draft section.

Try to write in note form – you don't have time to write sentences.

Important: The writing test is a test of your ability to extract relevant points from the stimulus material (whether they are medically correct or not) and write a letter.

Assume that you will comment on 2 or 3 issues and complete the letter by asking for something to be done.

Hint:
Start a notebook to learn the phrases which come up again and again in your writing. Phrases such as
* **undergo an operation /procedure**
* **make a referral**
* **need education on the use of (a peak flow meter/glucometer etc.)**
* **commence/start on a new medication.**

You can add to the notebook all the time. The more you review the words, the better.

It is also important to review the verb tenses which are often used in writing, e.g. the passive voice.
For example,
1. *Mr Blogs was reviewed by the gastroenterologist on September 14* is more common than *The Gastroenterologist reviewed Mr Blogs on September 14.*

2. *Mrs Smith was commenced on a daily diuretic yesterday.*

The OET Writing for Nursing sub-test takes 45 minutes. It is profession specific. You take this part of the OET using materials specifically for your profession i.e. a nurse does the task for nursing, a dentist does the task for dentistry, and so on. The test presents a task based on a typical workplace situation.

The OET uses a type of assessment which uses a simulated situation of an authentic situation. An example of this type of assessment is the OSCE (or OSCA) – Objective Structured Clinical Examination (or Assessment). By placing the candidates in a simulated situation it makes the assessment standard.

In addition, candidates do not have to be already working in the field so they can rehearse a common workplace task before performing it in the real world.

The most important skills you need to do the referral letter are:
* good organisational skills to put the stimulus notes in order
* ability to scan for relevant pieces of information
* ability to form 3 to 4 paragraphs of no more than 200 words in total

* ability to set out the address, date, salutation and ending correctly

What kind of the Tasks are used in the Subtest?

There are several types of letters which you may be asked to write:

1. a letter of referral, which asks for something to be done, e.g. a dressing

2. an informational letter, which gives advice, e.g. Quit Smoking or which informs, e.g. how to change a colostomy bag.

A lot of the information which you will need in your letter can be found in the discharge plan. It is important to only select the information from the discharge plan that is relevant to the person you are writing to.

With the task instructions, you will also receive stimulus material (case notes and/or other related documentation). This is the information which is used in your response.

Some of the information is not directly relevant to your letter. You have to decide what needs to go into the referral letter and what needs to be omitted. Remember that in real life, the patient notes and medication list will be sent along with the referral letter. So, the referral letter is to ask for something specific to be done. Some background is important but only if it is needed.

The first five minutes of the test is reading time. During this time, you may study the task and notes but <u>may not write, underline or make any notes</u>.

When you read the stimulus material, ask yourself, 'What is relevant to the person I am making the referral to?' It's a good idea to look quickly at the referral task (at the bottom of the task) before you start reading the stimulus material. Then you can read the stimulus material to see which parts are relevant.

You write your response to the task in the answer booklet during the remaining 40 minutes. This also has space for rough work. You are allowed to write in pen or pencil.

What Skills are used in the Writing Sub-test?

Take a moment to think about the skills you need to develop before doing the OET writing test. Think about what skills you need more practice in.

1. The OET tests you on your ability to use language which is specific to the medical environment.

How do you rate your understanding of medical-specific language in English?

Remember that the test is predominantly a test of your ability to communicate in English, however, a knowledge of medical terms is still needed to be successful in the test.

Improve your knowledge of medical terminology:

* read medical articles on reputable websites, e.g. Medscape, Medpage Today, medical journals etc.

* read articles about topics which relate to the 12 professions assessed by OET. Remember that the reading and listening tasks may cover topics from Veterinary Science, Podiatry, Occupational Therapy, Dentistry as well as Medicine and Nursing.

* ensure that you are familiar with pharmacology language – names of common drugs and types of medication (e.g. anti-hypertensives). Medication should be given as the drug name rather than brand name, e.g. Paracetamol rather than Panadol.

2. You will need to understand acronyms and abbreviations used in medical notes.

How do you rate your understanding of these terms in English?

Be careful when learning abbreviations. Note that American abbreviations sometimes differ from Aus/NZ/Brit terms. E.g. (American) EKG – (Australian) ECG. Also be aware that some abbreviations have been removed from use because they were considered unsafe e.g. (old) U – current 'Units' (of insulin.

When you are writing your letter, try to avoid the overuse of abbreviations or acronyms. As a guide, use no more than 2-3 abbreviations or acronyms in each letter. Whilst the letter is an example of a professional letter, it is also an English language test – a test of your ability to write clear, cohesive sentences.

This is an example of where an OET referral letter differs from an authentic referral to a medical colleague. In an authentic letter, healthcare professionals may use quite a lot of abbreviations or acronyms.

In the OET referral letter, you should aim to use a couple of acronyms or abbreviations to show that you know how to incorporate them in a letter in a meaningful way. For instance, it would make more sense to write '*The patient needs to have his INR checked next week'*, rather than *'The patient needs to have his International Normalised Ratio checked next week.'*

On the other hand, sometimes you should explain acronyms/abbreviations in the referral letter in a way that shows you know what they mean. For instance, if the stimulus material says that a medication has been increased from 'bd' to 'tds', you could write that the patient is now taking the medication three times a day rather than twice a day.

Improve your knowledge of abbreviations and acronyms:

* ensure that you know acronyms/abbreviations **used in Australia** for drug times (e.g. qid is used in Australia rather than qds).

* Learn commonly used acronyms/ abbreviations for cardiac (e.g. AF, ST elevation, HT, BP), respiratory (sats, COPD, CPAP, SOB), diabetes (bsl or bgl more common in Australia than cbs or BM (UK terms). Check that you know the commonly used acronyms/abbreviations for each body system/ dentition.

* learn expressions to use in place of acronyms/abbreviations or how to incorporate acronyms/abbreviations into a sentence, e.g. *He was given a stat dose of Frusemide.*

3. The OET writing task is to write a letter of referral e.g. to a community nurse, nursing home, hostel or GP.

How do you rate your ability to write a formal letter in English?

Do you understand how to set out a formal letter and close the letter correctly?

Important points to remember when writing formal letters/emails:

* **NEVER use SMS/text language.** You will lose marks if you write:

U (instead of *you)*

Cld (instead of *could)*

Wld (instead of (*would)*

pl (instead of *please)*

i (instead of (*I)*

im (instead of *(I'm)*   etc.

* You only have to learn how to set out a letter correctly once, then just copy it! Become familiar with the use of Australian postcodes, i.e. *SUBURB + STATE + POSTCODE,* e.g.

Dulwich Hill   NSW   2203

If no postcode is given, leave it out. Just write as much of the address as you can, e.g.

Mr John Silver

Grace Nursing Home

Deepwater

* Write the address using an Australian format, i.e.

DAY / MONTH / YEAR - 3/9/2015 is the 3rd of September. Be careful writing the date in this format, if you are used to the American system (MONTH / DAY/ YEAR). The above date would be the 9th of March in this case.

* The correct salutation is one of these options:

Dear Sir, (if you know it is a male you are writing to and you are being very polite)

Dear Mr Silver, (for a male who you do not know well or at all). **Never write 'Dear Mr John,' – this is incorrect.**

Dear John, (for a male colleague who you know well and are on first name terms with. This may be appropriate for colleague-to-colleague referral letters)

Dear Madam, (if you know it is a female you are writing to and you are being very polite) **Do not write 'Dear Mam' – it is incorrect.**

Dear Ms Silver, (in Australia it is more common to use 'Ms' for both 'Miss' and 'Mrs', unless the person has specified otherwise. For instance, if a letter is signed:
(Mrs) Jenny Summer – address her as 'Dear Mrs Summer,' or
(Miss) Susan Orwell- address her as 'Dear Miss Orwell,'

Dear Sir/Madam, - this is used, if you do not know who you are writing to. For example, the stimulus material may say to write to the Community Nurses at Deepwater Community Health Centre.

DO NOT USE THIS EXPRESSION:
'Hello dear,' - it is informal and very old fashioned.

4. An important part of the task is to read 'stimulus' material which is usually in the form of a patient discharge summary.

How do you rate your ability to scan for relevant information?

Improve your ability to scan for relevant information

* before you start reading a text, predict what you think you might find in the text. For instance, a text about dental caries in children may well include:
dental caries
tooth/teeth brushing
sugar
gum disease/gingivitis
tooth extraction

* the stimulus material is set out under headings. Firstly, look at the age of the patient. If the patient is elderly, you can probably expect to find some of the illnesses of old age or chronic illnesses, e.g. arthritis, dementia (perhaps). A patient who has had a car accident may well have broken bones.

Hint: the OET may use common illnesses, such as diabetes, in the stimulus materials, however you will find that there will be a complication or related issue. For instance, a teenager with diabetes who also suffers from depression.

Expect to find information about at least two medical conditions. Try to link information about medication with the conditions, e.g. the patient may have changed medication/ take an increased dose of a drug etc.

* develop the skill of identifying the **key terms** in a sentence or chunk of information. For example, in the following stimulus material the terms in bold are the terms you should keep in mind:

hypertension 5 yrs
*recent change of anti-hypertensives – now **well controlled needs monitoring** regularly – can forget to take tabs*

* as you read the stimulus material, think about where you will put the information, i.e. which paragraph.

5. During the 5 minute reading time, you are not allowed to make any notes or underline any information in the stimulus information.

How do you rate your ability to remember the information you need or to remember where to find the information?

6. Certain grammatical forms, e.g. passive forms are often used in formal letter writing. Note that you will mostly use simple sentence structures which should be grammatically correct.

How do you rate your use of correct grammatical structures?

How to use grammatical structures confidently?

Learn a few common phrases which can be modified when needed, e.g.
*He/She **has had** (hypertension, arthritis, etc.) **since 2006.***
*He/She **has had** (back pain etc.) **for a week /six months / two years.***
*He/ She **was monitored / was assessed / was reviewed by…***

Typical Assessment Types for Medical Professionals

There are three main types of assessment which can be used to assess the performance of tasks. These are:

1. Observation of the person performing tasks at work over a specified time. An example of this are the regular staff appraisals which healthcare workers are subject to on an annual basis.

2. Assessment of the performance of specific tasks performed at work. An example of this may be the assessment of a nurse preparing and giving an IV injection. This type of assessment may be called competency-based assessment.

3. Assessment of the performance of specific tasks in a simulated situation. An example of this are the OSCEs (or OSCAs) –Objective Structured Clinical Examination (or Assessment).

The Occupational English Test uses the third type of assessment to check the English language skills of a health care professional. It is useful to place candidates in a simulated situation because it standardises the assessment. In addition, candidates do not have to be already working in the field and also allows them to rehearse a common workplace task before performing it in the workplace.

How is the OET marked and assessed?

Firstly, it is important to remember that your work will be marked by two experienced assessors who work independently. This means each assessor approaches your paper on his/her own. Your paper will be analysed using FACETS software to make sure that any differences in marking difficulty or leniency by particular assessors is picked up. If there is any doubt, a third assessor is brought in.

The assessors will give a score from 1 (lowest) to 6 (highest) for each of the five criteria. They have a detailed guide to help them decide the mark. **Remember that, unlike IELTS, the interlocutor who plays the part of the patient or carer during the role play IS NOT AN ASSESSOR.**

After the initial analysis of the five criteria, any tests which have scores which do not fit the expected pattern are scored again by a third assessor without the assessor knowing the first score reached. The final score is then reached.

Criterion	Description
Task fulfilment	* Have you followed the instructions of the task? * Are you writing the correct type of letter? (referral to a nursing home, follow-up letter to existing nursing home, letter to the GP requesting an action) *Is the letter the correct word length?
Use of appropriate language	* Did you use formal language rather than 'chatty' language which is more appropriate for a letter to a friend? * Did you use medical terminology where necessary? * Is the tone of the letter non-judgemental? e.g. *'The patient has difficulties with compliance'* is better than *'The patient doesn't try to take his medication at all.'*

Understanding of the stimulus material	* Did you scan the patient notes (stimulus material) for relevant facts? * Did you understand abbreviations and symbols used? e.g.↑ SOB (increasing shortness of breath) * Did you only use relevant facts in your answer? That is, facts which go under the paragraph headings of your letter.
Use of correct grammar and text cohesion	* Can you make sentences using correct grammar forms? Note that the passive is often used in formal letters. * Do your sentences make sense (have cohesion)?
Use of correct spelling and punctuation	* Do you know how to use commas and full stops? * Can you spell common terms used in medical English?

What Level of English are you at?

The minimum pass mark for OET is B for all four subtests. A B at OET is the equivalent of IELTS 7 – 7.5 or C (Advanced) in the CEFR (Common European Framework of References of Languages).

Most language schools who offer specialised tuition in OET only offer classes to candidates who are at a minimum of B2 level (Upper Intermediate).

Look at the following chart and try to assess where you are **at the moment.** This will give you an idea of how much study and preparation is needed before you attempt the OET.

OET	IELTS	CEFR	Description
A	8-9	C2	**Mastery: you can** * easily understand with ease almost everything you hear or read * summarise information from

			different spoken and written sources and present an argument coherently

* express yourself very fluently and precisely. You can differentiate shades of meaning, even in the most complex situations |
| B | 7-7.5

Nurses 7

Doctors 7.5 | C1 | **Advanced: you can:**
* understand a wide range of complex texts, even if the topic is unfamiliar

* express yourself fluently and spontaneously without much obvious searching for expressions, even if the topic is unfamiliar

* use appropriate language for social and professional purposes

* produce well-structured and detailed written texts on a range of complex subjects

* use complex grammatical forms, such as connectors and cohesive devices |

C	6-6.5	B2	**Upper Intermediate: you can** *understand the main ideas of complex texts, including technical discussions in your field of specialisation or when the topic is familiar * speak quite fluently and spontaneously on familiar topics, but find it difficult to cope with unfamiliar topics * write clear and detailed texts on familiar subjects and explain a viewpoint on a topical issue, e.g. the advantages and disadvantages of a viewpoint.
D	5-5.5		
E	4-5	B1	**Intermediate: you can** * understand the main points of a conversation about familiar topics * manage most situations during daily interactions * produce simple connected text on topics which are familiar or of personal interest and which use high frequency words

			* describe experiences and events and briefly give reasons and explanations for opinions and plans

As you can see, the main difference between a score of C in the OET and a score of B is the degree to which you can manage spoken and written texts about topics which are unfamiliar.

At C Level (B2 or Upper Intermediate), you should be able to manage texts and conversations about topics which you are familiar with. This means that you know a range of words and terms which are needed to be fluent. For instance, if you are talking about a hobby you have, you will have learned the related vocabulary, so that you can discuss the hobby easily.

If we think about a conversation of a medical topic, e.g. during a role play, you may know the related terms and understand the procedure in your own language, but not be familiar with the equivalent terms in English. If you struggle to cope with texts about unfamiliar topics, it may suggest that you are at C level.

In simple terms, OET C is Upper Intermediate Level and OET B is Advanced Level.

How Important is the Use of Good Grammar?

Writing sentences that are grammatically accurate is important but not as important as **getting your message across.** It is more important to be able to explain what needs to be done clearly than simply use the correct verb tense.

HINT: Keep in the back of your mind the reason why you are writing the letter.

Remember that most of your sentences will be quite simple so you can concentrate on writing clear and accurate sentences. You are not using descriptive language in most cases; you are passing on information and facts. It is better to write short accurate sentences rather than long confusing ones!

There are some grammatical structures which are useful to learn and practise.

1. Review verb forms used to indicate present time (what the patient is doing now), past time (what happened to the patient in hospital) and future time (what needs to happen for the patient in the next days and weeks). The passive is also used

quite frequently in formal writing.

Present time – often used to talk about the patient's current condition. Also used to talked about present habits with:

* *usually*
* *often*
* *sometimes*

For example:
Mr Davis is being transferred to Mount Gold Hospital today.
He had a total hip replacement last week.
Mr Smith mobilises with a stick / a wheelie walker / crutches
Mr Smith is unsteady on his feet.
She is chair bound / bedbound.
She is independent with her ADLs / independent with her personal care.
He has hypertension and high cholesterol.

Present Perfect – when a time factor is mentioned, e.g.
* *for*
* *since*

For instance:

for four months, since 1983.
He has had COPD since 2008.
She has had a leg ulcer for six months.
He has lived alone for approximately two years, since the death of his wife.

Past time – to explain treatment whilst in hospital. Also, time expressions using *ago*.

Mr X underwent a right knee arthroscopy.
Mrs Y underwent cardiac monitoring in our unit over the past three days.
Mrs X came to the hospital for sleep studies this week.
Mrs Smith has been discharged today.
He underwent a total hip replacement on the 19th March.
She underwent treatment for cancer of the pancreas.
He noticed a loss of hearing a week ago.

Future time – to talk about what the patient needs after leaving hospital. For instance:

Mr X will need to have his clips /sutures /staples removed on 5.4.2013.
Mrs Y will need to have a FBC and INR three days after

discharge.
Mr X is for repeat MRSA swabbing next month.
He will need his sutures removed in 7 days.
He will need his clips removed in 8 days.
She will need a dressing change in 3 days.
She will need an INR in a week.

Passive –The passive tense is often used in writing as it avoids clumsy sentences. Compare these sentences:

1. *He was recommenced on warfarin 2mg daily.* (less clumsy)
2. *The doctors recommenced him on warfarin 2mg today.* (clumsy)

For instance:
He was reviewed by the physiotherapist and given an exercise programme to do at home.
Her vital signs were monitored frequently after the episode of chest pain.

Look at the following examples of the passive voice.
He was commenced on warfarin 2mg daily.
She was recommenced on her anti-hypertensives today.

His digoxin was withheld for two days because his heart rate dropped.

He was diagnosed with anaemia.

He was found to have a low sodium level which was replaced with IV fluids.

He was transfused with two units of whole blood.

He was referred to a rheumatologist whilst an inpatient.

She was transferred to ICU for a few days after an episode of severe chest pain.

Use of Formal and Informal Verbs

When speaking, it is more common to use informal verbs, which may be phrasal verbs. In writing, try to use formal verb forms.

For instance:

Formal	Informal
assess	check out
commence	start
implement	put something into place
monitor	keep an eye on
recommence	start again
review	look at again
assist	help
perform (a procedure)	do

Verb and Preposition Pairs

Become aware of verb and preposition pairs. Note that some verbs have several 'preposition pairs' with different meanings.

For instance:
to care for: to look after
to care about: to have a positive feeling about a person or thing

Here is a list of common verb and preposition pairs:

care for	He cares for his grandmother.
monitor for	Could you monitor him for signs of infection.
be continent of	She is continent of urine.
be incontinent of	He is incontinent of urine and faeces.
be on	She is on twice daily insulin.
commence on	He was commenced on warfarin whilst in hospital.
recommence on	She has recommenced anticoagulant therapy.
be disoriented to	She was disoriented to time and

	place.
refer to	She was referred to the Diabetes Nurse.
transfer to	She was transferred to CCU immediately after her operation.
be compliant with	Unfortunately, he was not compliant with his medication.
continue with	She should continue with her post-op physiotherapy.
diagnose with	He was diagnosed with throat cancer two months ago.
be independent with	She is independent with her ADLs / personal care.
be dependent on	He is dependent on his wife for all personal care.
liaise with	Could you please liaise with the patient's GP about her warfarin dose?
transfuse with	He was transfused with 6 units of packed cells.
mobilise with	She mobilises with a walking stick.

Agreement of Verbs

Make sure that your verbs agree with the **subject** of the sentence. The subject is the 'doer of the action (verb)'. In the following sentence, *Mr Netherwood now takes an increased dose of diuretics:*

the action (verb) = take
the doer of the action (subject) = Mr Netherwood

What does 'agree with the subject' mean?

You need to match a single noun with the third person singular verb form (has, underwent, shows) and a plural noun with the third person plural with the third person plural form (have, show) e.g.
- *The X-ray shows a small area of consolidation in the right lung. (*Subject – X-ray; verb – show)
- *Both the CT scan and ultrasound indicated a small lesion in the liver. (*Subject – CT scan and ultrasound; verb – showed)
- *He has lived in the country all his life. (*Subject – he; verb – lived)

There is / There are

Make sure that you use the expressions *There is* and *There are* correctly.

There is + singular noun

There are + a plural noun

He or She?

Be careful to use 'he' for males and 'she' for females. If your language makes no distinction between male and female, be especially careful.

Parallel Sentences

If you want to make your sentences more complex, you can join several ideas in one sentence. This is called a **parallel sentence**.

Parallel sentences often join the ideas using:
- as
- and
- or

Make sure that parallel sentences are balanced properly.

1. Join using 'so'

*During the peri-operative period, Mr Hooper experienced a significant amount of pain, **so** a PCA was commenced.*

2. Make sure that passive forms agree. In other words, do not mix active and passive forms in the same sentence.

For example:

*During the post-operative period, anti-emetics **were given** for nausea and vomiting and IV fluids **were administered** to avoid dehydration.*

3. There is no need to repeat 'was' in the past passive, e.g. was transfused....(was) commenced

*His haemoglobin level decreased rapidly, so he **was transfused** with three units of blood and **commenced** on iron supplements.*

4. Make sure that gerund forms agree. A *gerund* is a verb form which ends in 'ing' and acts like a noun to name an activity. Some examples are:
Mobilising with a walking stick makes walking safer for many elderly people.
Chewing slowly and carefully is recommended after patients have a stroke.

*Mr Hooper developed chest pain on the second post-operative day when **mobilising** or **having** a shower.*

5. Make sure nouns and noun phrases agree.

*Her blood glucose levels continued to rise over the first two post-operative days because of a high **carbohydrate intake** and the **administration** of steroids.*

Become Familiar with Common Drugs

Medication Used in Heart Disease

Here is a list of medication commonly used to treat heart disease. Make sure you are familiar with the types of medication used in Australia and common examples.

ACE Inhibitors – dilate arteries to lower blood pressure	Lisinopril (5mg, 10mg, 20mg) Ramipril (1.25 mg, 2.5 mg, 5 mg, 10 mg)
anti-arrhythmia drugs	digoxin (62.5 mcg, 250 mcg)
anti-coagulants	warfarin (1mg, 2mg, 5mg)
anti-platelet drugs – blood thinning drugs	Aspirin 100 mg chlopidogrel 5mg
beta blockers – block the effect of adrenaline	Atenolol 50 mg Bisprolol 1.25 mg, 2.5 mg, 5 mg, 10 mg
Diuretics – removal of	Frusemide 20 mg, 40mg,

excess water from kidneys	500 mg Bumetanide 1 mg Spironolactone 25mg, 100mg
statins	Atorvastatin 10 mg, 20mg, 40mg, 80 mg Simvastatin 5 mg, 10 mg, 20 mg, 40 mg, 80 mg
vasodilators – nitrates	GTN spray 400 mcg, 600 mcg GTN patch (5mg/24hrs, 10mg/24hrs)

Practice Examples:

OET Writing for Nurses: Peripheral Artery Disease

Case Notes:

Mrs Martha Sculthorpe, 75 years old, is a patient in the medical ward of which you are Charge Nurse.

Hospital: Dartmore Public Hospital, 61 Main Street, Dartmore

Patient details

Name: Martha Sculthorpe (Mrs)   **DOB:** 16/6/1940

Marital status: Widowed

Address: 12 Roseville Ave, Seaview 2366

Next of kin: daughter, Fiona – works part-time, visits twice a wk.

Admission date: 14/5/2015

Discharge date: 24/5/2015

Diagnosis: chronic, infected arterial ulcer upper right foot

Past medical history:

Peripheral Artery Disease 10 yrs.

Intermittent claudication - pain in calf after walking > 10 m
Hypertension 12 years - takes Verapamil 40 mg daily
Smoker- 5 cigs a day >50 years. Unwilling to quit smoking, becomes breathless easily

Social background: lives in own home. Daughter supportive but does not live close.
Not coping well at home – accepts Homecare for cleaning but refuses MOW

Admission Notes:
Admitted with infected arterial leg ulcer – community nurse noted redness around ulcer, incl. discharge and offensive smell, increased pain in right foot causing ↓ mobility.
On adm. T 38.2, general malaise.

Tests and Imaging:
U/S (Doppler) of Right lower leg – mod. narrowing of arteries in lower leg.
Wound swab 15/5 - bacterial infection

Nursing Management:

Daily alginate dressings to ulcer. Compression bandage applied daily after dressing.

Cephalexin p.o. for 10 days.

Quit Smoking programme discussed with patient - agreed to cut down but feels unable to stop smoking altogether.

Encouraged to elevate leg when sitting and to use walking frame when mobilising.

Assessment at discharge:

Afebrile.

Wound bed cleared of infected exudate.

↓ pain when mobilising – ind. with walking stick

supervise personal care – needs assistance showering

Discharge Plan:

1) Assist with personal care and reapplication of compression bandage

2) Redress wound in 7 days.

3) Review at Dressings Clinic in 2 weeks' time - appt for 9 June to be mailed to pt.

3) Monitor compliance with Quit Smoking programme, encourage pt to continue.

Writing Task:

Using the information given in the case notes, write a discharge letter to Ms Elizabeth Clarke, Community Nurse at the Community Health Centre, 10 Lupus Street, Seaview 2366.

Planning the Referral Letter

1. The referral letter is to be written to a colleague (a community nurse – Ms Elizabeth Clarke). All relevant details have been provided for the address and salutation. Add the details to the template you always use.

2. Look at the discharge plan – what are you asking the community nurse to do? Answer: These four points in the discharge plan.

a) help with personal care – you can link this with the social history. The patient is elderly, lives alone, has a daughter who does not live close by and who can only assist a few days a week. In addition, the patient has accepted some assistance

(Home Care), but has refused Meals on Wheels.

b) continuing wound management – ensuring compression bandage applied every day, redressing wound and ensuring that the patient returns to the clinic for dressing review.

c) monitoring of smoking cessation

Therefore, there are three main issues which need to be addressed in the early part of the referral letter.

Paragraph 1: Introduce the background of the patient, including mention of her chronic leg ulcer and smoking habit. Explain the patient's social background (inability to cope at home).

Paragraph 2: Explain the management of the leg ulcer and the attempts made to encourage the patient to quit smoking.

Paragraph 3: Make requests of the community nurse (following the discharge plan).

School Nurse Letters: Letter to a Parent

The next example is a type of letter which a school nurse writes to a parent. Typically, the case notes will document one or two visits to the school nurse. One or two problems will have been identified, possibly after the child's teacher has reported the child to the school nurse for assessment.

One type of letter is written to the parent – an issue is explained and treatment is suggested, e.g. the use of lice shampoo or cream for impetigo ('school sores').

A second type of letter is one which a school nurse writes to a GP to report a health issue. The health issue may have been identified during a previous visit to the nurse. In addition, there may have been recommendations made which have not been followed.

The next set of case notes relate to a letter sent by the school nurse to a child's parent.

Case Notes: School Nurse and Childhood Obesity

You are the school nurse at Seaview Primary School. A 10-year-old girl, Gisela Volkman, has been sent to see you by her teacher because of concerns about Gisela's unwillingness to join in with regular class-room activities.

Patient Details
Name: Gisela Volkman
DOB: 14 January 2004
Aged: 10
Parents: Peter and Sabine Volkman
Home address: 16 French Park View Rd, Seaview 7356

Social Background
Gisela is the middle child (siblings all attend Seaview Primary School) of the family - only child in family with JA
Family very close.
Tend to be protective of Gisela bec. of chronic health issues

Medical History

Current Weight: 47 kg (above 97th percentile)

Current Height: 138 cm (50th percentile)

Eczema as a baby- occasional flare-ups treated with 0.5% Betnovate cream (max 1 week)

Juvenile idiopathic arthritis - diagnosed age 8. Mainly affects knee and ankle joints.

Mild asthma - uses salbutamol inhale prn

Recent Medical History

2014 March   Referral by class teacher. Gisela unable to manage physical activity during PE or at playtime because of pain in knees and ankles.

General health check – wt. 40kg, ht. 137cm. Teacher reports seeing Gisela eat large morning break snacks and large lunches as well as additional food purchased from the Tuck Shop.

Reports that Gisela not keen to join in PE activities such as swimming claiming that 'my mother says I should rest as much as I can'. Phoned mother and advised review of pain medication by GP to enable Gisela to join PE activities.

Offered referral to Child dietician for guidance on diet but mother declined. Feels she can manage at present.

2014 July: Teacher referred Gisela to me again today. Pain in knees and ankles now controlled by current analgesia, however, Gisela still unwilling to join into physical activities.

Weight gain noted (now 47 kg). Teacher reports that Gisela has become more withdrawn in the past two months and appears embarrassed by her weight gain. Apparently, she has been avoiding weekly swimming lessons because of this.

Gisela is keen to speak to a dietician now. She would also like to attend a swimming programme away from school; advised about hydrotherapy programme at Summerfield Hospital.

Phoned mother to discuss today's visit- no answer to phone call so letter sent home with Gisela.

Medication
Ibuprofen 600mg qid
Salbutamol inhaler qid and prn

Writing Task

Using the information in the case notes, write a letter to Gisela's mother outlining discussion with Gisela about her weight gain and reduced activity level.

Suggest appointment with paediatric dietician and offer to refer if acceptable.

Also inform about availability of hydrotherapy programme at Summerfield Hospital for arthritis sufferers. Offer to refer Gisela to the programme.

Explain importance of exercise programme in management of arthritis.

In your letter:
Expand the relevant case notes into complete sentences
Do not use note form
Write between 180 - 200 words
Use correct letter format

Planning the letter

During this type of letter, it is important to use non-judgemental language when writing to the mother.

Judgemental Language

Examples of judgemental language include:

* references to patients refusing to follow instructions, e.g. *'Patient was advised to give up smoking, however, has refused to make any attempts to quit or cut down her smoking habit.'*

* linking a patient's ethnic background with negative stereotypes, e.g. *'Patient is of Aboriginal background and has a high alcohol intake.'*

* references to marital status, e.g. *'Patient is a single mother.'* It would, however, be relevant to state that a patient lives alone and does not have a strong support network.

The instructions at the end of the case notes help you to make a plan of the letter you will write to the mother. It is clear that you have to write about:

a) weight gain – effect on the child's arthritis as well as reduced activity level

b) child's interest in swimming – beneficial for arthritis as well as child's body image.

c) concerns about obesity – tactful offer of dietician review

School Nurse Referral to another Medical Professional OET Writing for Nurses: Teenager with Anaemia

You are Francie Clooney, the school nurse at St Mark's School for Girls. Read the case notes and write a letter of referral.

In this case, the nurse is writing to a medical colleague, so it will be appropriate to use medical terminology, rather than everyday terms.

Case Notes:

Patient details
Elspeth Simmonds
36 Kenmore Rd
Tamswood 3267

DOB: 4 /2 / 1999

Presenting complaint:
Heavy periods

tiredness

? iron deficiency anaemia

Social background

Lives with mother and younger brother (5 years old)

Parents separated, father lives interstate

Mother works as a teachers' aide

Vegetarian for past 2 years

Medical history

2008   removal of mole from lower back (benign)

2011   onset of menarche
            dysmenorrhoea (especially first and third days).
            often absent from school on first day of period.
            takes Paracetamol 1g qid

2012   periods quite heavy
            last up to a week on average
            menstrual cramps continue – GP changed prescription to Mefenamic Acid 500 mg tds and prn for menstrual cramps.

Recent history

May 2014 – Elspeth came to see me about heavy periods. Changes tampon and pad every 2 hours on first and second days. Discussed diet - advised iron-rich food (she is vegetarian) - sent patient information leaflet home for mother about avoiding anaemia on vegetarian diet.

August 2014 – Elspeth came to see me again today - complaining of extreme tiredness. No sleep problems reported. Sleeps at least 10 hrs. per day. Has slight breathlessness and lower back pain. Elspeth still suffers from heavy periods which have become more irregular. Advises that she has made no dietary changes.

Writing Task

Using the information in the case notes, write a letter of referral to Elspeth's GP, Dr Stephen Campbell. Ask the doctor to review Elspeth's condition and advise on treatment options. Also suggest that a dietician assessment may be helpful. Address the letter to: Dr Stephen Campbell, Health Clinic, Samsford 3265

In your answer:

- Expand the relevant case notes into complete sentences
- Do not use note form
- Use letter form
- The body of the letter should be approx. 180 - 200 words.

Planning the referral letter

The case notes contain some information which is not relevant for the referral letter. Also, some of the information may be construed as judgemental, e.g. 'Parents separated, father lives interstate' and 'Advises that has made no dietary changes.'

As you write the letter, it is important to merely report that no dietary changes have been made. Do not imply that that the student has not been trying to make changes to improve her health status.

Planning the letter

Because you are writing to a medical professional, it would be appropriate to write a referral letter which is similar to hospital type referral letters.

In this case, you do not have a definite diagnosis (it is 'possible iron deficiency anaemia'), which you, as a nurse cannot diagnose. This is one of the reasons for the referral to the GP – to assess the condition and advise on treatment.

Paragraph 1: Sets out the background of the health issues, i.e. possible iron deficiency anaemia caused by heavy periods and vegetarian diet which is not supplying sufficient iron to the student's diet.

Paragraph 2: Explain the nursing interventions you have already put in place, i.e. advice on iron-rich foods suitable for vegetarians. Also, report your observations to the GP (heavy blood loss during periods, reports of tiredness)

Paragraph 3: Request for review of student and advice on further treatment.

Case Notes: Deliberate Self Poisoning

This is an example of a Mental Health referral letter. Note the comments about the patient's mother – *'Mother does not believe in anti-psychotic medication. Very religious, believes in power of prayer.'* If you make any reference to the part his mother plays in the patient's recovery, you must be culturally sensitive.

What is Cultural Sensitivity?

Cultural sensitivity is a knowledge and awareness of a patient's ethnicity, culture, gender or sexual orientation, so that the patient's responses to their health environment can be better understood.

Cultural sensitivity presupposes an acceptance of different cultural responses to medical care and treatment.

Case Notes

Mr Wallace Gribic is a 28-year-old male patient on the Mental Health Unit at St Alban's Hospital where you are a Charge Nurse.

Name: Wallace Gribic
Age: 28
Admission Date: 5/4/2015
Discharge Date: 26/4/2015

Diagnosis: Deliberate Self-Poisoning (DSP)

Sociocultural background:
Serbian, has lived in Australia since a child (5yrs old)
Divorced 1yr ago - depressed since then.
Lives with mother who speaks little English
Mother does not believe in anti-psychotic medication. Very religious, believes in power of prayer.
Pt inactive – does not exercise at all and tends to spend a lot of time playing video games

Current medical history:

Paranoid Schizophrenia - diagnosed 3 yrs. ago

2nd episode of deliberate self-poisoning

Past Medical History

Depressive Illness 6 mths ago after recent divorce

1st episode DSP also after divorce

2009   peptic ulcer – treated cimetidine

2010   # L clavicle 2010 (involved in fight)

Medications:

Clozapine 100 mg daily – reluctant to take, as does not like side-effects (constipation and increased salivation)

Senna prn (constipation)

Inpatient Summary:

* stabilised medically after DSP

* counselling with mental health worker – identified triggers which increase depression, introduced coping strategies.

* commenced CBT

* commenced Clozapine – educated about side effects

Advised to increase fibre in diet and take laxatives when necessary for constipation. Advised chew gum to help swallow saliva. Needs supervision to ensure takes meds.

* family meeting with mother and psychiatrist – aim to explain benefits of medication. Mother still not convinced about the medication.

Discharge Plan:

1) Start case management via community Mental Health Team, include mother in family discussions

2) Continue to reinforce importance of anti-psychotic meds (benefits, side effects management)

3) Continue CBT

4) Encourage ↑ exercise – persuade pt of benefits of exercise and benefits of ↓time spent on video games.

Writing Task

You are the Charge Nurse on the Mental Health Unit where Mr Wallace Gribic will be discharged from. You need to write a nursing referral letter to the local Community Mental Health Team. Address the letter to Ms Flora Bates, Team Leader, Rosemont Community Mental Health Team. 17 Vesper St, Rosemont NSW 2317.

Use the discharge plan notes to refer your patient for community follow up.

In your answer:

- Expand the relevant notes into complete sentences
- Do not use note form
- Use letter format
- The body of the letter should be approx. 180-200 words

Planning the letter

This letter has two examples of why a patient may not be compliant with medication:

1) The patient's mother is suspicious of anti-psychotic

medication, believing instead in the power of prayer. As the patient lives with his mother, it will be important to include the patient's mother in team meetings about the patient's continuing care.

2) the patient is reluctant about taking the anti-psychotic medication because of the unpleasant side-effects he has experienced. Because of this, there is a recommendation that the patient take measures to minimise these side-effects (constipation and dry mouth).

You will notice that the discharge plan in this case is substantial. The referral letter should therefore concentrate on the patient's care on discharge from hospital.

The first part of the letter should cover the reason for admission to the Mental Health Unit (second DSP) as well as a brief medical and social background of the patient.

The second part of the letter should incorporate recent medical treatment into the request for follow up care:
1) encourage use of anti-psychotics and reinforce ways to minimise side-effects. Add need to educate patient's

mother on the value of taking anti-psychotics, whilst respecting her cultural view.

2) ask for continuation of CBT, mention success of inpatient counselling.

3) encourage increase in exercise – for both physical and mental health benefits.